Maths
Foundation Plus
Anthology

T0187088

Published by Collins
An imprint of HarperCollins*Publishers*
The News Building, 1 London Bridge Street,
London, SE1 9GF, UK

HarperCollins Publishers
1st Floor, Watermarque Building, Ringsend Road,
Dublin 4, Ireland

Browse the complete Collins catalogue at
www.collins.co.uk

10 9 8 7 6 5 4 3 2 1

ISBN 978-0-00-846890-3

British Library Cataloguing-in-Publication Data
A catalogue record for this publication is available from the British Library.

Compiled by: Peter Clarke
Publisher: Elaine Higgleton
Product manager: Letitia Luff
Commissioning editor: Rachel Houghton
Edited by: Sally Hillyer
Editorial management: Oriel Square
Cover designer: Kevin Robbins
Cover illustrations: Jouve India Pvt Ltd.
Additional text credit: Hannah Hirst-Dunton
Internal illustrations: p 2–5 Laura Gonzales, p 6–11 Amanda Enright, p 12–13,
22–23, 30–31 Priya Kuriyan, p 14–15, 26–27 Tasneem Amiruddin, p 16–19 Jouve
India Pvt. Ltd., p 20–21 Q2A Media, p 24–25 Brett Hudson, p 28–29 Mike Phillips
Typesetter: Jouve India Pvt. Ltd.
Production controller: Lyndsey Rogers
Printed and Bound in the UK using 100% Renewable
Electricity at Martins the Printers

Acknowledgements

With thanks to all the kindergarten staff and their schools around the world who
have helped with the development of this course, by sharing insights and
commenting on and testing sample materials:

Calcutta International School: Sharmila Majumdar, Mrs Pratima Nayar, Preeti
Roychoudhury, Tinku Yadav, Lakshmi Khanna, Mousumi Guha, Radhika Dhanuka,
Archana Tiwari, Urmita Das; Gateway College (Sri Lanka): Kousala Benedict; Hawar
International School: Kareen Barakat, Shahla Mohammed, Jennah Hussain; Manthan
International School: Shalini Reddy; Monterey Pre-Primary: Adina Oram; Prometheus
School: Aneesha Sahni, Deepa Nanda; Pragyanam School: Monika Sachdev; Rosary
Sisters High School: Samar Sabat, Sireen Freij, Hiba Mousa; Solitaire Global School:
Devi Nimmagadda; United Charter Schools (UCS): Tabassum Murtaza; Vietnam
Australia International School: Holly Simpson

The publishers wish to thank the following for permission to reproduce photographs.

p 16–19 Shutterstock

MIX
Paper from
responsible sources
FSC™ C007454

FSC
www.fsc.org

This book is produced from independently certified FSC™
paper to ensure responsible forest management.

For more information visit:
www.harpercollins.co.uk/green

The pond

I spider, I duck, I butterfly,
How many tadpoles do you spy?

Look quickly at the groups you see.
Which have 9? Which have 3?

Rav's bedroom

10 crayons on the bed, 10 teddies in the bed!
Is there the same number of books to be read?

Across the room are zooming cars.
Count up and write how many
there are!

Along the river

Jack trots along behind his sister Jill,
Over the river and up the big hill.

The men on the water, inside their small boat,
Are near some green frogs on a log where they float.

What number of sheep at the farm can you spy?
Is it more than the number of birds in the sky?

Look at the fields. What's the number of trees?
Is it smaller than 9? Is it bigger than 3?

In town

The castle is standing opposite town.
The cart goes on the path between
them, up and down!

High above the castle, how many
flags show?
Are there fewer than the number of
windows below?

Minibeasts

How many caterpillars can you spy?
How many worms and butterflies?

Which groups can you see have a
total of 4?
Which groups have fewer? Which
groups have more?

Necklaces

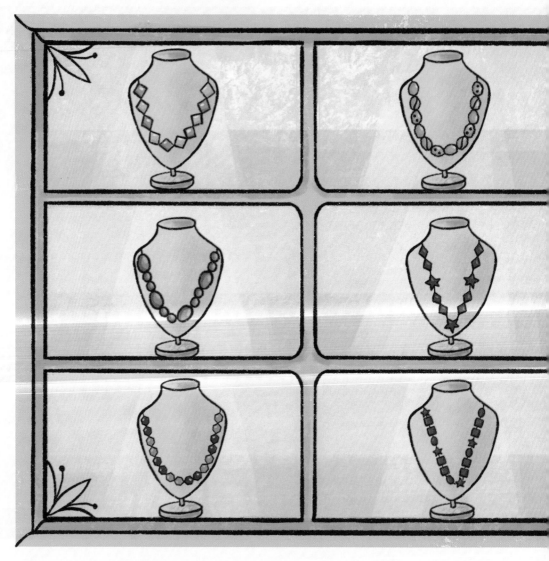

See the necklaces these beads make,
And all the beads' colours, sizes
and shapes.

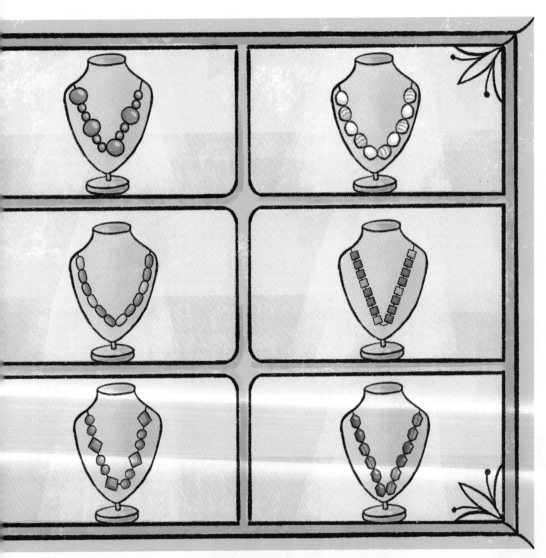

Each necklace has a pattern of beads. To make one longer, which beads would you need?

What different things can circles be?
How many rectangles can you see?

Find a triangle, find a square.
Are there any more shapes there?

3D shapes

Cylinders, cones, spheres and cubes,
pyramids and cuboids, too.

Find shapes that roll and shapes that stack. Did you spot a few?

Jenny's classroom

Ted keeps his bag behind him, on the corner of his chair.

Kim leans on the edge of the table with his hand up in the air.

The bookshelf's near the toy box.
Outside of the window's a tree.

The crayons are under my best
friend's hand — and next to her is me!

The zoo

Which is the tallest giraffe in the zoo?
Which is the heavier kangaroo?
Look for the shortest crocodile, too!

Which lion has the widest mane?
Which pathway is the narrowest lane?
Which animals are all the same?

The see-saw

Look at the animals on the see-saw! Which weigh less, which weigh more?

How many animals came to play?
Who joined the fun? Who ran away?

These are the vases in Yasmine's store.
Some hold less and some hold more.

Which vases are empty in the shop?
Which are full, right to the top?

At 9 o' clock, where is Sven?
Can you see what he did then?

At lunch, what did the clock face say?
What happened last, to end the day?

The toy cupboard

Look at the toys in this display!
Why have groups been made
this way?

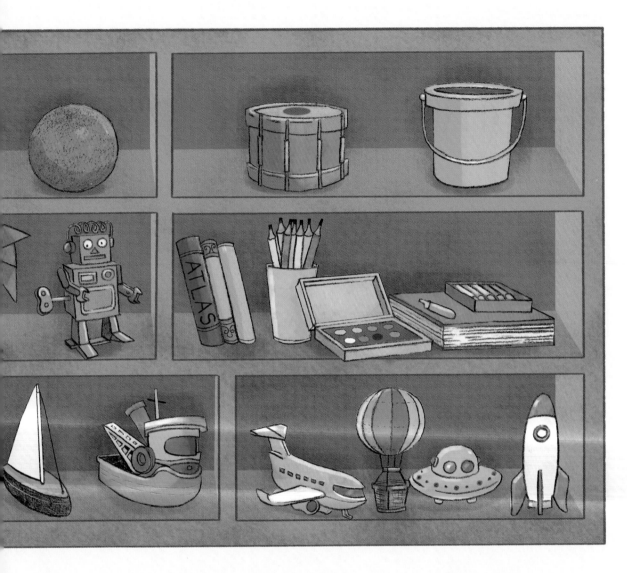

Do some toys fly? Do any float?
Find the spheres, then find a boat.

Numbers 0 to 10

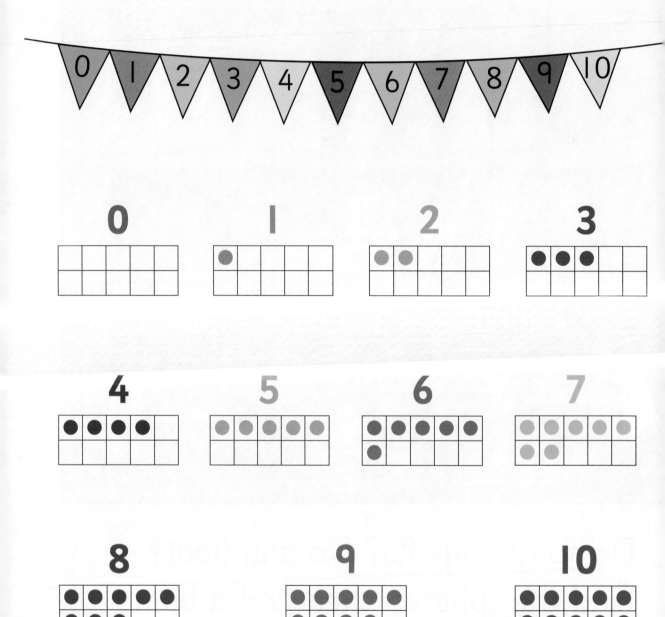